Effect of Ladanum Gnawing on Enterostasis Following Caecal Helminthoid Extirpation

Valentinus

Effect of Ladanum Gnawing on Enterostasis Following Caecal Helminthoid Extirpation

First Edition August 2025

Written by Valentinus

TABLE OF CONTENTS

INTRODUCTION

Appendicectomy is a widely performed surgical procedure worldwide. The prevalence of appendicectomy has increased in the past few decades owing to a multitude of variables, such as a poor diet, particularly among the younger population who lead lives of motionlessness.

Due to this specific factor, the problems and incidence related to appendicectomy, whether performed by open surgery or laparoscopy, are now on the rise.

Up to 5% of patients are said to experience post-operative paralytic ileus, which results in stomach distension, vomiting that causes the patient tremendous distress, and a lengthy hospital stay that necessitates a number of procedures. Patient satisfaction scores have decreased, there is a greater need for painkillers, and as a result, healthcare costs have climbed.

With the implementation of ERAS procedures in surgery, early feeding schedules have received special attention. However, very few doctors are reluctant to begin oral feeding at a young age. Additionally, 20% or more of patients cannot tolerate early feeding.

Gum chewing alone increases saliva production and stimulates the muscles of the stomach and intestines as well as the neuro-humoral system. By boosting the stomach's motor activity, this in turn promotes quicker gastric emptying.

Better patient satisfaction, a decline in the need for painkillers, faster hospital discharge, and lower treatment costs have all resulted from this.

Gum chewing is becoming increasingly favoured by many institutions and is included in "Enhanced Recovery After Surgery" protocols because it can give the advantages.

Although the risks of aspiration and increased stomach capacity have been presented theoretically, past research have revealed that these risks are merely theoretical and do not actually translate into difficulties. With this context, it has been suggested that the study who have had appendicectomy.

This study aims to prove that sham feeding through chewing gum can prove to increase gastro-intestinal motility in turn reducing the occurrence of post operative ileus hence ensuring a better and quick post operative recovery.

REVIEW OF LITERATURE

APPENDIX

A small, blind-ended tube called the appendix is connected to the posteromedial end of the cecum.

Despite having a significant amount of lymphoid tissue, it is not believed to play any crucial roles in the human body.

The posteromedial region of the cecum is where the appendix develops. The meso-appendix, a fold of mesentery that lets the appendix hang from the terminal ileum, provides support for it.

Depending on how it connects to the ileum, caecum, or pelvis, the position of the free end of the appendix can be divided into seven basic regions. The retrocecal posture is the most typical. They might also be identified by how they appeared in reference to a clock:

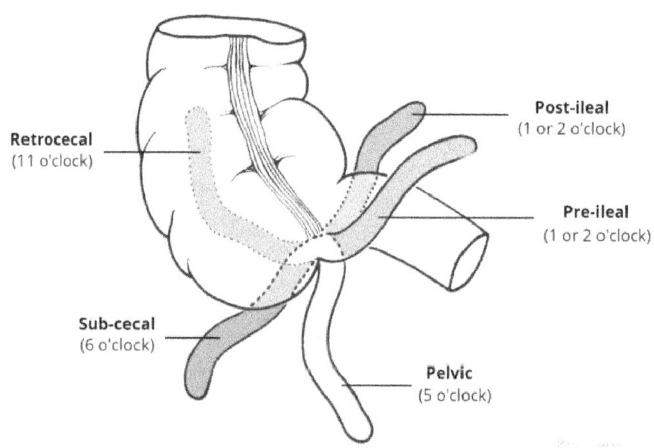

FIGURE 1 : The different positions of appendix (Teach Me Anatomy)

The embryological midgut gives rise to the appendix. As a result, branches of the superior mesenteric arteries provide the vascular supply.

The autonomic nervous system's sympathetic and parasympathetic ganglia innervate the appendix. The superior mesenteric plexus' ileocolic branch helps to do this. To get to the appendix, it travels down the ileocolic artery.

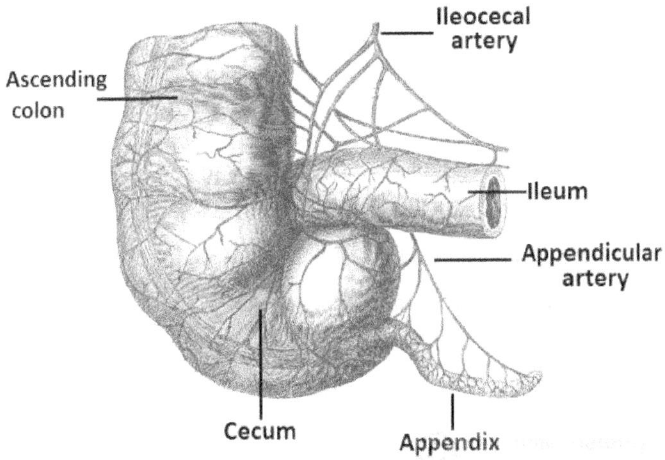

FIGURE 2 : Arterial Supply to the Cecum and appendix via the ileocecal artery (Teach Me Anatomy)

Both the ileocolic lymph nodes (which round the ileocolic artery) and the

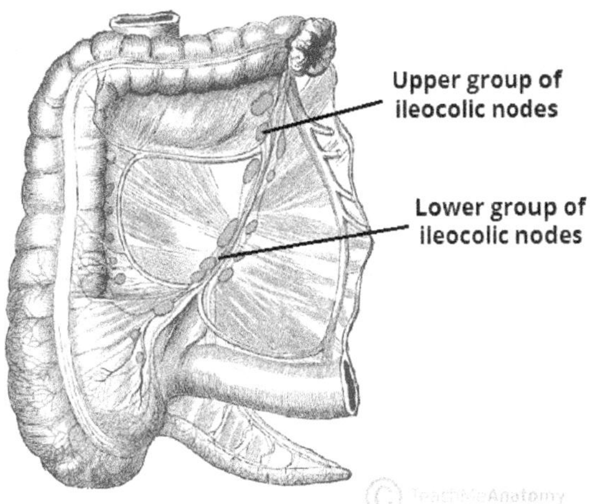

mesoappendix receive lymphatic fluid from the appendix.

FIGURE 3 : Lymph nodes of appendix (Teach Me Anatomy)

APPENDICECTOMY

HISTORY OF APPENDICECTOMY

In September 1731, English surgeon William Cookesley performed the first appendectomy that has been documented as being successful. The second occurred on December 6, 1735 , at S t. George's Hospital in London, where surgeon Claudius Amyand noted a perforated appendix inside an 11-year-old boy's inguinal hernial sac. The organ appeared to have been pierced by a pin the youngster had ingested. Charles McBurney formulated the theory that a n early appendectomy was the best course of action to prevent perforation & peritonitis in 1889 in New York City after precisely describing the presentation and aetiology of appendicitis.

There have been some instances of autoappendectomies. Evan O'Neill Kane tried one in 1921, but his assistance finished the procedure. Another was Leonid Ro gozov, who had to operate on himself in 1961 since he was the sole medical professional on a remote base.

Kurt Semm performed the first laparoscopic appendectomy on September 13, 1980, paving the way for a much wider use of minimally invasive surgery.

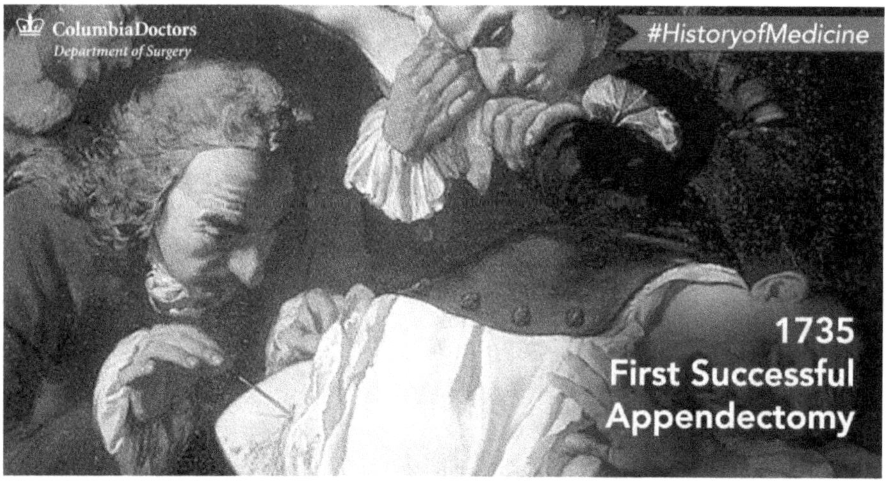

FIGURE 4 : Picture depicting the first appendicectomy done in the St.Georges Hospital

INDICATIONS

Presently appendicectomy has been strongly recommended in patients with the following indications :

- Clinical appendicitis : where in the patient comes in with an acute classical presentation of right lower abdominal pain , fever , tachycardia.
- Uncomplicated & Complicated appendicitis
- Diverticulosis of appendix (rare)
- Crohn disease appendix (rare)
- Appendiceal tumors (rare) : mostly a postoperative finding

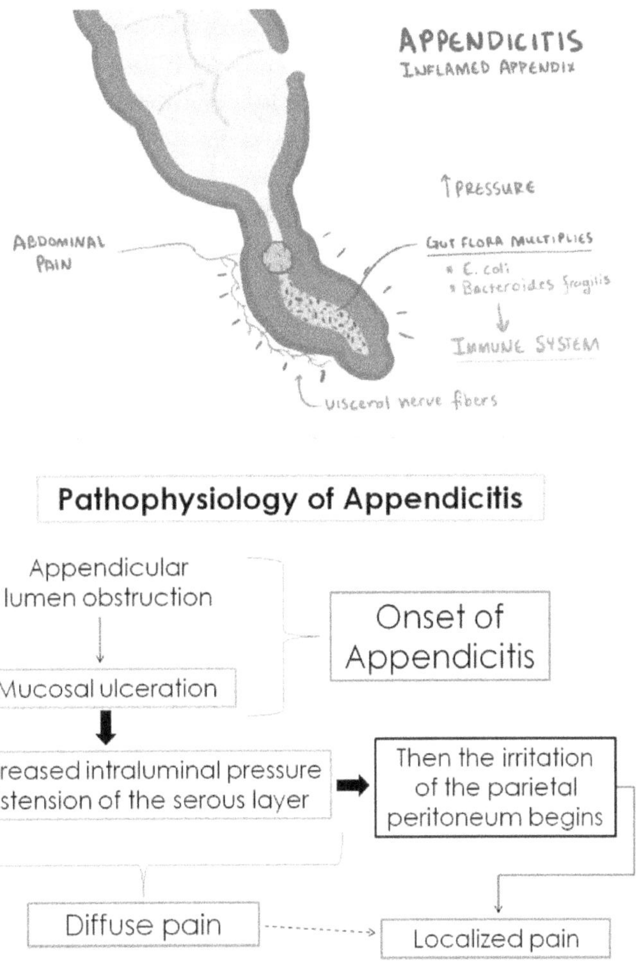

FIGURE 5 : Pathophysiology of Appendicitis

TECHNIQUE:

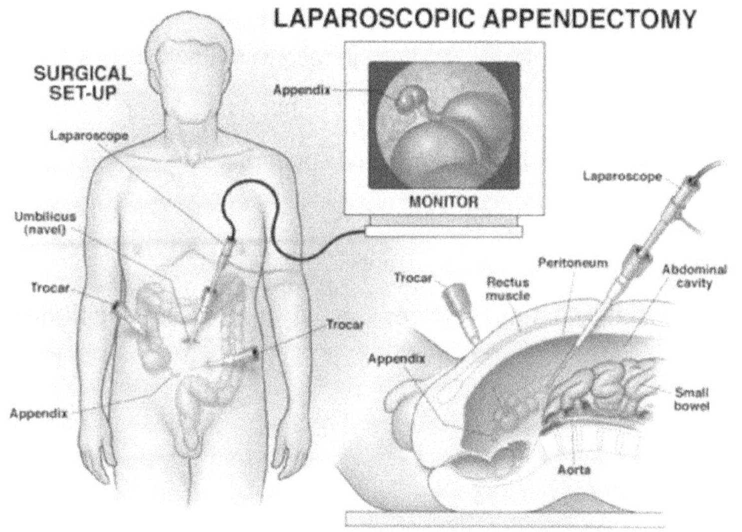

Laparoscopic Appendectomy

FIGURE 6 : Picture depicting port position in laparoscopic appendicectomy (SAGES)

- Making a 12 mm infraumbilical incision allows for access to the abdomen, which is then reached by utilising either the Hasson procedure or a Veress needle to inflate carbon dioxide gas into the peritoneal cavity.
- Next, a 5-m m angled laparoscope is put in. A second 5 mm port i s then positioned laterally in the left lower quadrant after the first 5 mm hole is positioned midline over the pubic bone.
- The lateral port is used to insert the laparoscope. The midline ports will be used during surgery by the surgeon.
- The abdomen is then examined to rule out any further illnesses. The right lower quadrant is now the focus of attention.
- The patient c an now b e placed into a steep Trendelenburg with their rig ht side u p to help with exposure.
- The small bowel and omentum are then both swept cephalad. The terminal ileum can be located using the fold of Treves (antimesenteric fat).

- The terminal ileum is followed in order to locate the cecum. The appendix can then usually be found by following the cecum's taenia.
- Frequently, adhesions between the appendix (vestigeal organ) , terminal ileum mainly , and cecum are seen; these adhesions can be separated via blunt dissection.
- Sharp dissection might be required, though. Notably, electrocautery should not be used when operating close to the bowel to avoid contact and conductive harm.
- It should be thought about switching to an open appendectomy if visualisation or dissection are challenging.
- Once the appendix's tip has been identified, it can be gripped and raised anteriorly off the cecum.
- A window is made between the meso-appendix and the appendiceal base using a Maryland dissector.
- The appendix and mesoappendix should be positioned parallel l to the lateral side abdominal wall to minimise the risk of bowel damage and ensure the safest orientation for this window. The mesoappendix is separated with an endoscopic stapler loaded with vascular staples.
- T he appendix is divided at its base with the endoscopic stapler and the second load of vascular staples. To avoid leaving an appendiceal stump, care mus t be given to ensure that the stapler is brought together close to the cecum.
- After that, the excised appendix is put in an endoscopic retrieval bag and taken out using the supraumbilical port.
- The integrity and hemostasis of the mesoappendiceal and appendiceal staple lines are then determined.
- The lower abdomen receives the small bowel after it has been cleaned caudad. Ideally, the appendiceal stump should also be covered by the omentum, which is pulled back over the small bowel.
- The laparoscope is placed through one of the midline ports to directly view the removal of the remaining ports and to evaluate the anterior abdominal wall's hemostasis.
- When opposed to fascial closure, port sites may be left open, which lowers the chance of a needle stick, the time it takes to complete, and the overall cost of the surgery.

- The skin is subsequently stitched up using monofilament sutures, skin adhesives, or staples as the surgeon sees fit.

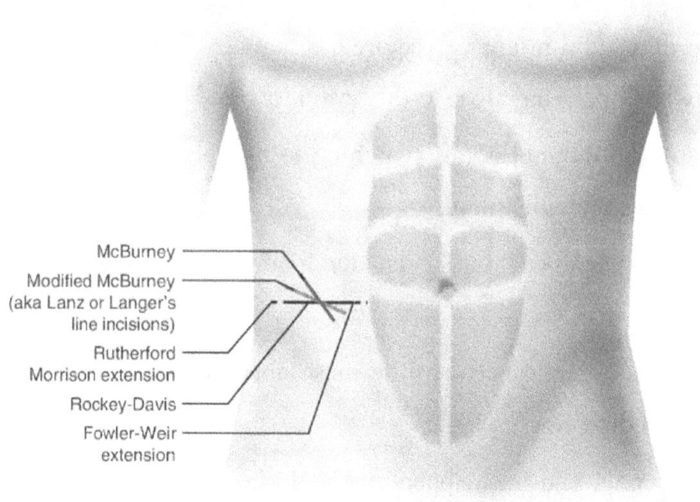

Open Appendectomy

FIGURE 7 : DIFFERENT INCISIONS FOR OPEN APPENDICECTOMY (SPRINGER)

- On the basis of preoperative imaging, an incision can also be performed at the site of greatest discomfort or the appendix.

- A scalpel is then used to make an incision in the epidermis and dermis. To dissect the external aponeurosis, blunt dissection and electrocautery are employed.

- The internal oblique muscle beneath the external aponeurosis is then exposed by opening the aponeurosis from superolateral to inferomedially along its fibers.

- The internal oblique is then split bluntly perpendicular to the course of its fibers to expose the trans verse abdominal muscle.

- The peritoneum i s subsequently exposed by utilizing blunt dissection to separate the transverse abdominal muscle in a manner that is parallel to the direction of its fibers.

- The peritoneum is grabbed with forceps and cut with a 15-blade scalpel. After that, the appendix's location comes into focus.

- The cecum can be moved and used to locate the appendix if it can be seen.

- The appendix can be externalized by advancing with the taenia coli while using atraumatic graspers.

- Alternatively, by sweeping a finger around the cecum from a supero-lateral position and moving infero-medially to locate the appendix, the cecum can be utilized as a reference to localizing the adding.

- After locating the appendix, the mesoappendix should be removed, and the appendiceal arteries should be divided into clamps.

- Sutures made of silk are then used to ligate the appendiceal vessels. The base of the appendix can then be sutured using a silk purse-string.

- The appendiceal base tissue is broken up using a right-angle clamp. The clamp is then pushed distally, and the crushed tissue is where the appendiceal base is ligated using an absorbable suture.

- After that, the appendix is removed just in front of the right angle clamp using a 15-blade knife. Electrocautery can be used to completely remove the appendiceal stump mucosa.
- Then, using a silk purse-string suture, invert the stump.
- Confirming hemostasis is necessary.

- The peritoneum and fascial layers are stitched shut using absorbable sutures.

- The skin is stitched shut with absorbable sutures or how the surgeon sees fit.

COMPLICATIONS

- Due to the fact that appendectomy is a reasonably safe treatment, the death rate is minimal.

- Geographical differences in appendectomies' mortality rates—more industrialized nations having rates of 0.09 to 0.24 percent and underdeveloped nations having rates of 1 to 4 percent—have been shown.

- Mortality and morbidity rise in cases of appendiceal perforation and are correlated with disease stage.

- However, due to delayed diagnosis, the perforation rate is substantially higher in small children and older individuals, where it has been calculated to be up to 97%.

- There is an 8.2 to 31.4 percent total complication rate after an appendectomy, a 9.4 percent pelvic abscess rate, and a 3.3 to 10.3 percent wound infection rate.

POSTOPERATIVE ILEUS

PATHOGENESIS OF POSTOPERATIVE ILEUS

T o have a thorough grasp of the anatomical pathology be hind post-operative para lytic ileus, one must be familiar with the physiology and normal stomach motor function.

There are three different ways that gastrointestinal processes are controlled :

- Sympathetic and Parasympathetic enteric nerves

- Cajal's interstitial cells and enteric neurons

- Smooth muscle cells in the gut

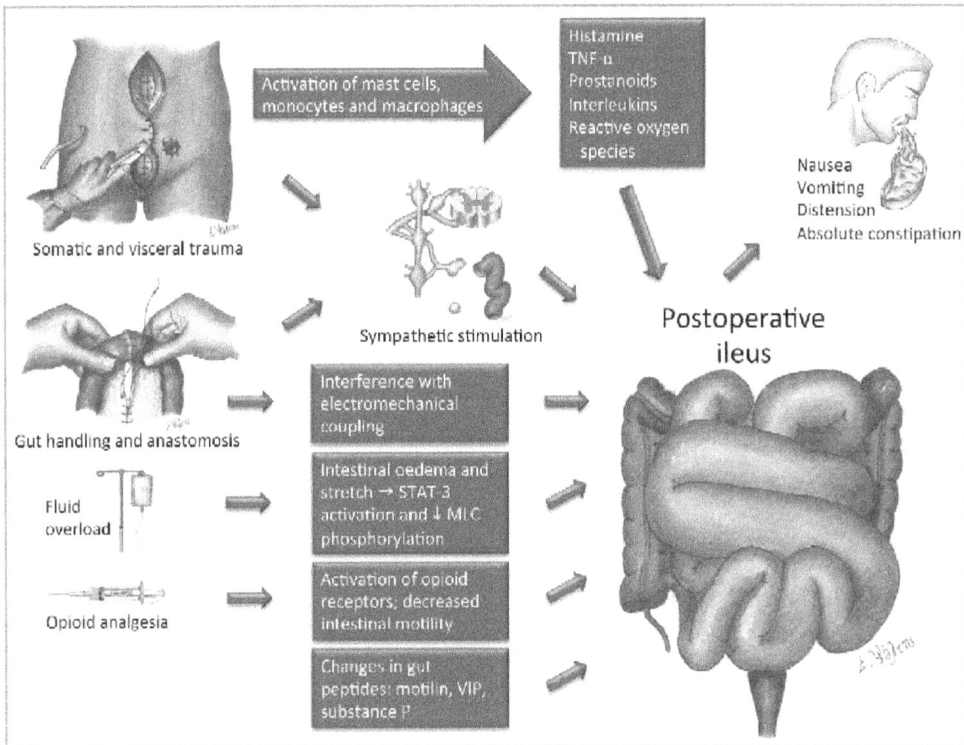

FIGURE 8 : Pathogenesis of Post operative ileus

AUTONOMIC NERVOUS SYSTEM

The spinal segments T5 to T10 supply the stomach with sympathetic energy. The celiac ganglia are where they exit the spinal cord after emerging from the inter mediolateral columns of the aforementioned spinal cord segments.

SMOOTH MUSCLE

The primary motor unit of the digestive system is made up of smooth muscle cells. The cell membranes of smooth muscle cells have certain receptors that can bind to particular compounds. The smooth muscle cells get these neurotransmitters by a variety of endocrine, paracrine, or neurocrine pathways.

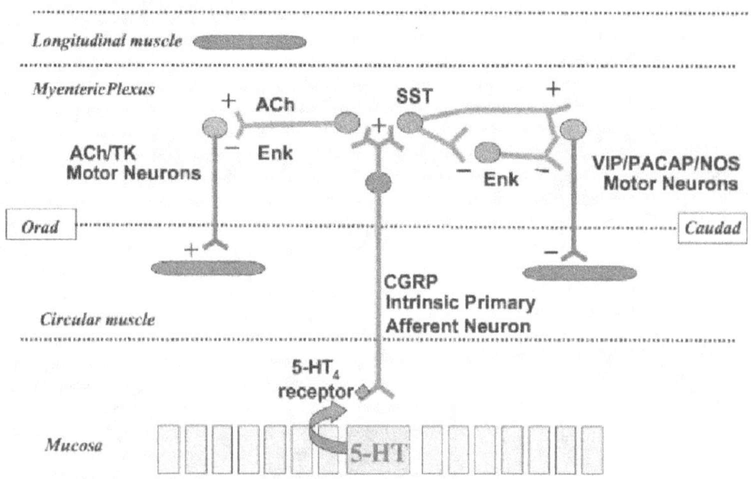

ROLE OF HORMONES

Pharmacological methods are utilized to manipulate GI hormones externally, physiologically replace them, and produce their agonists.

THE PERISTALTIC REFLEX

Food is propelled from the stomach into the intestine as a result of the peristaltic reflex. There are two elements that make the meal move forward.

FIGURE 9 : Hormones acting on Peristaltic reflex

MECHANICS OF DIGESTION IN STOMACH

The musculature of stomach has three layers

- Circular layer
- Longitudinal layer
- Oblique layer

The two primary functioning components of the stomach are the fundus and antrum. The pacemaker for the gastric electrical function is located in the centre of the stomach's greater curvature. The "housekeeper," or migratory motor complex, is what moves the indigestible wastes to the colon. MMC happens while one is fasting and helps to clear

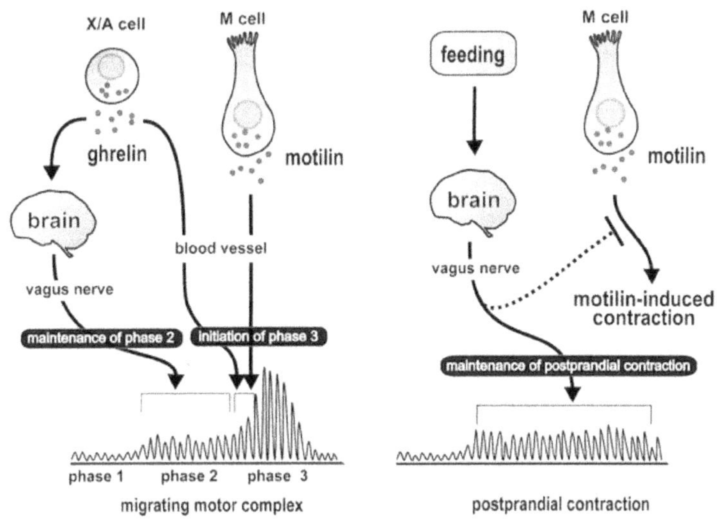

the digestive system. The meal is then poured into the antrum after that.

FIGURE 10 : Migratory motor complex

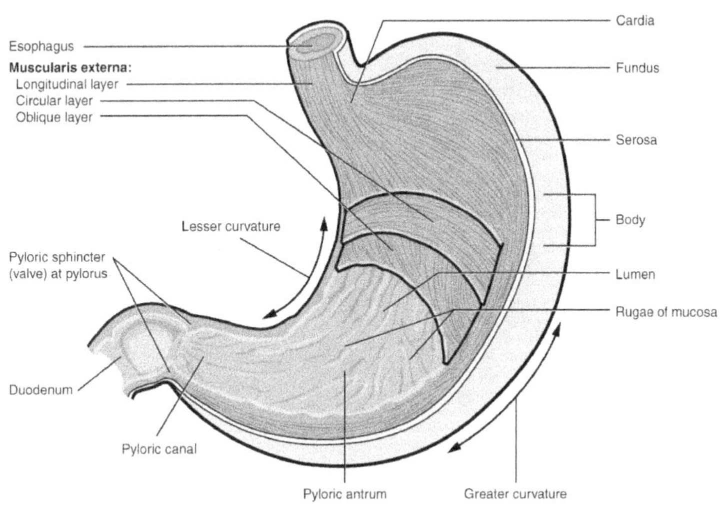

FIGURE 11 : Picture depicting Stomach musculature

THE ANTRUM

The stomach's antrum is characterized by strong contractions that exert shearing stresses on the body. Trituration is the procedure used to decrease solid food particles to a size of 1 to 2 mm. Larger sol id food particles are repeatedly driven and retro-pulsed to produce trituration in t he blocked distal stomach.

The antral motor function is principally responsible for the grinding, mixing, and stomach emptying motions. The antral motility significantly correlates with the rate of liquid evacuation from the stomach. To enable trituration, gastric motility is programmed to occur with a defined lag time. The interdigestive antral motor function empty the undigested solid matter in between meals.

Gravitation and the distal part's fall in resistance both contribute to the emptying of liquids, which can also be a passive process. A zone of high resting pressure is the pyrorus. According to radiological measurements, it is between 0.6 and 1.6 centimeters long. Three phasic contractions take place per minute in opposition to a closed pylorus. Acetylcholine and nitric oxide in turn govern these contractions.

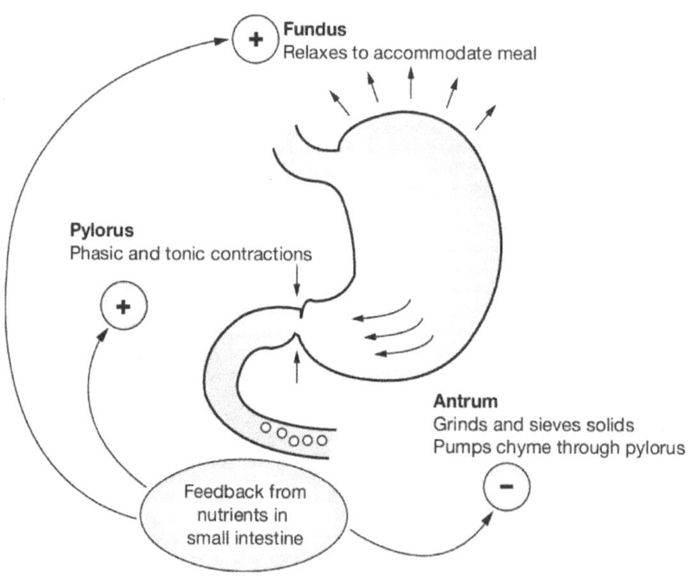

FIGURE 12 : Picture depicting Antral peristalsis

PARALYTIC ILEUS

According to estimates, 8.5% of routine abdominal procedures result in paralytic ileus.

Two forms of paralytic ileus are possible.

MANIFESTATIONS

There are no set standards or criteria for what constitutes paralytic ileus. The length of ileus varies depending on the procedure and technique, therefore it is unclear at what point it becomes prolonged.

Paralytic ileus is typically considered protracted if it lasts more than three days. The following are symptoms of paralytic ileus:

- Abdominal kneading / distension
- Abdominal pain
- Vomit and nauseous
- Belching
- Delayed exit of stool or flatus
- Hiccups
- Anorexia

RESOLUTION

Normal bowel function returning becomes a crucial requirement for patient discharge. Thus, the timing of the recovery of bowel function is important.

Although it has been discovered to be incredibly inaccurate, the restoration of bowel sounds in any quadrant i s typically the first indication of normal intestinal function.

PERSISTENT ILEUS

Recovery is slowed down by paralytic ileus, which also causes nausea, vomiting, abdominal pain, and a greater need for medications. As a result, enteral feeding is delayed, and additional issues such atelectasis, aspiration, intestinal bacterial translocation, nosocomial infections, dehiscence, and venous thromboembolism arise.

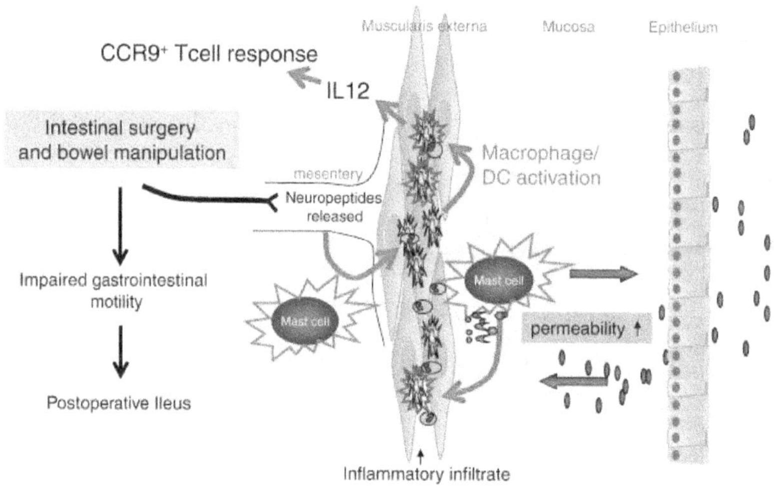

FIGURE 13 : Inflammatory Reactions promoting post operative ileus

CAUSES OF ILEUS

- MEDICATIONS
- METABOLIC CAUSES
- SYSTEMIC ILLNESS
- INFECTIONS
- NEUROLOGICAL
- AUTOIMMUNE

Surgery-related stress raises sympathetic tone and inhibits the release of acetylcholine via raising catecholamine levels. An increase in sympathetic activity is brought on by lengthy procedures and adhesiolysis, which then inhibits bowel movement..

SHAM FEEDING

Sham feeding refers to the practice of simulating normal food consumption while without really swallowing, digesting, or absorbing the food. Gum chewing is seen.

These gastrointestinal neurons may be stimulated by the chewing. This process starts the release of digestive hormones by stimulating the cephalo-vagal system. Additionally, it causes an increase in saliva production and pancreatic and intestinal secretions, which speeds up stomach emptying.

As a result, sham feeding is being researched as a way to increase gastrointestinal movement. The biggest benefit of early oral feeding is that it has no negative effects, especially for those who do not take it well. 20% of patients can't eat without getting sick. Gum chewing is a low-cost, physiologically safe technique that is linked to earlier bowel motility, appetite restoration, bowel noises, flatus passage, and stool motions.

In past research, only minor adverse were documented. This has led to increased preference among institutions and inclusion in "Enhanced Recovery After Surgery" protocols, as well as higher patient satisfaction, reduced use of painkillers, quicker discharge from the hospital, and lower treatment costs.

The effectiveness resulting from causes other than surgical ones has been investigated in a variety of research, including randomised control trials. Major procedures include colorectal, gynecological, urological, gastric, and hepatic surgeries have all had positive outcomes recorded. Gum chewing in the postoperative period has been found in numerous randomised control trials to result in less postoperative ileus, early recovery, and faster hospital discharge.

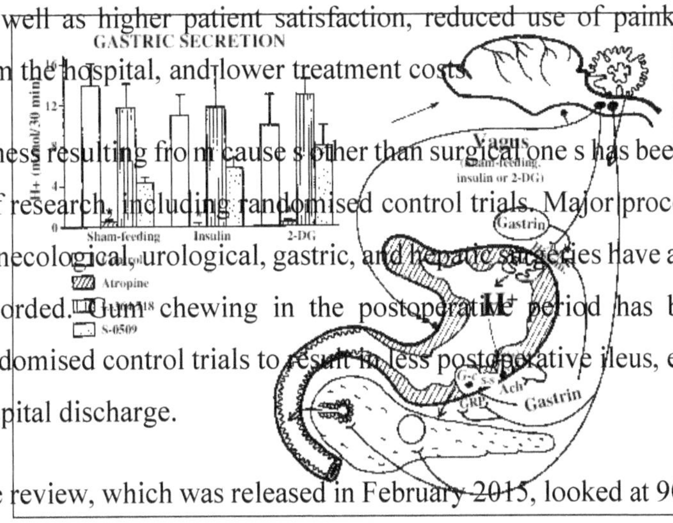

The Cochrane review, which was released in February 2015, looked at 9000 patients in 81 pertinent RCTs. The review was in favor of enhancing gastrointestinal motility by chewing gum. However, the research was primarily comprised of short, low-quality trials and primarily focused on colorectal procedures. In a comprehensive review and meta-analysis that Wenceslao Vasquez published, 6 studies with 244 patients were examined.

The amount of time it took for the first stoma to pass was also dramatically shortened. Gum chewing only slightly shortened the length of stay. And it was suggested to include chewing gum in the typical post-operative therapy regimen.

FIGURE 14 : Picture depicting SHAM Feeding and correlating acid secretion

They came to the conclusion that sham feeding was safe after colorectal surgery, and the GI healing had improved somewhat but noticeably. In a meta-analysis a nd s ystematic review of 9 e ligible trials including 4 3 7 patients, Emma J. Noble et al. found that elective intestinal operations generally resulted in better outcomes.

The intervention was accepted nicely. The time it took for these patients to pass their first flatus was reduced by 14 hours, the time it took for them to pass their stools was decreased by 23 hours, and they were discharged earlier by 1.1 days.

In a randomised control experiment published while 48% of the control group experienced post-operative ileus, only 27% of chewing gum users did. The chewing gum group's participants (85 vs. 87%) passed feces four days after surgery. P 0.006).

MATERIALS AND METHODS STUDY DESIGN

The study was based on an analysis of 60 patients with appendicitis who underwent treatment in CHRI, Kelambakkam. 60 patients have been equally distinguished into 2 divisions with 30 peoples assigned in each of the groups who were one managed by sham feeding and regular post operative care and two only by regular post operative care . Diagnosis was made on the basis of thorough history and clinical examination in the OPD.

INCLUSION CRITERIA

1. All post operative appendicectomy patients from Post operative day 0 to 3
2. Age group :15 to 50 years
3. Sex : both sexes

EXCLUSION CRITERIA

1. AGE GROUP: less than 15 and more than 50
2. Immuno compromised patients .
3. Patients on steroidal therapy .
4. Patients with diabetes , connective tissue disorder .
5. Patients with neuro muscular disorders
6. Patients with major electrolyte imbalance .

METHODOLOGY

- Study Design – Randomised Control Trial
- Sampling Technique – Randomised sampling
- Study Area - Chettinad hospital and research Institute, Kelambakkam. Tertiary care centre General Surgery Department
- Sample Size - 60 (30 : cases , 30 : control)
- Study Population - Patients with appendicitis

- Duration of study – 1 year (August 2021 to June 2022)

PROCEDURE:

- After taking proper consent from the patients in a language they understand best , all patients routine investigations were taken and assessed. They were made to understand about the procedure in detail in a language they understand best.
- **Procedure venue :** MAJOR Operation Theatre
- **Procedure position :** SUPINE POSITION
- The patients are given the routine pre procedure medications such as : - Inj Tetanus Toxoid 0.5 mg IM and test dose of the local anaesthetic agent (Inj Xylocaine 0.2cc intradermal) .
- Once the diagnosis is confirmed and the patient is planned for surgery , they are in detail explained about the study.
- Patients undergo appendicectomy under spinal or general anaesthesia based on open or laparoscopic approach.
- Once the procedure is over , the patients are kept in the SICU for the night for observation.
- The test group_will be receiving sugarless bubble gum from the point of full consciousness post operatively thrice a day and will be asked chew for a period of 20 min till bowl sounds are heard .
- The control group will follow the regular protocol as per the surgeons orders .
- Findings will be tabulated for evaluation (which usually includes initiation of bowel sounds, flatus and passage of stools) .
- All the findings were tabulated and then compared between both the groups.

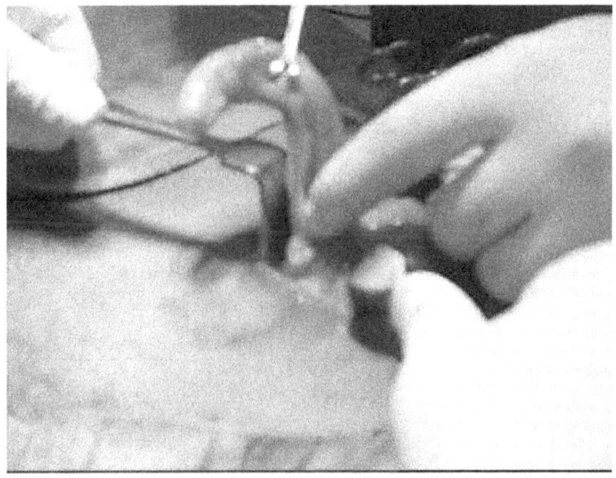

FIGURE 15 : Picture depicting Open Appendicectomy

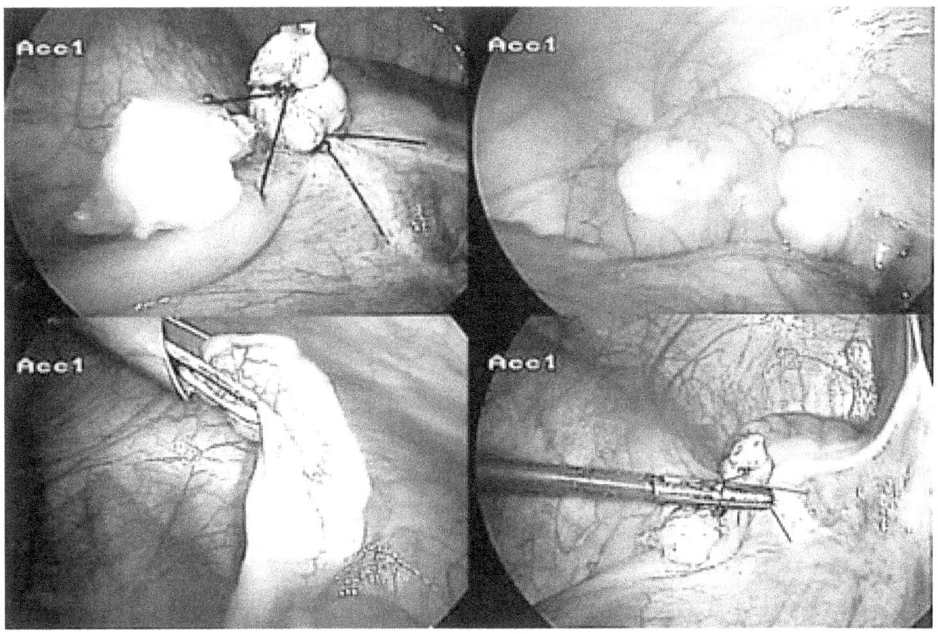

FIGURE 16 : Picture depicting Laparoscopic Appendicectomy

OBSERVATION AND RESULTS

AGE PREDOMINANCE

Table 1: Age distribution among study participants

Age Category	Group A		Group B	
	Number of Patients	Percentage (%)	Number of Patients	Percentage (%)
15 to 30 Years	19	63.3 %	21	70 %
> 31 to 60 years	11	36.7 %	9	30 %
Total	30	100%	30	100%
Minimum Age in years	15		16	
Maximum Age in years	60		56	
Mean Age in years ± SD	29.17 ± 11.97		28.07 ± 12.05	
t value	0.01			
p value	0.724			

Note p value based on Independent samples Unpaired t test

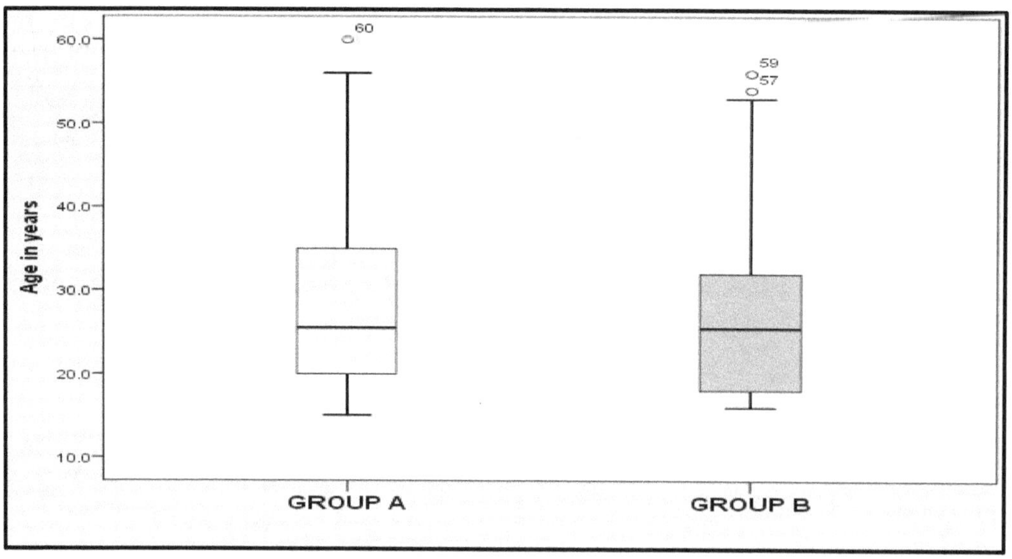

Graph 1 : Box whisker diagram shows age distribution among study participant

The average age of group A was 29.17 + 11.97 years while group B was 28.07 + 12.05 years. The two parties were identical in age (P=0.724). The youngest age limit was 15 in group A and 16 in group B. The maximum age was 60 in group A and 56 in group B. Most of them are 15–30 years old—63.3% in group A and 70% in group B.

Hence as statistically proven there isn't a statistical difference in age group between the two groups and also it is found that appendicitis is more common in the age group of 15 to 30 years of age.

SEX PREDOMINANCE

Table 2: Gender distribution among Study participants

Sex Distribution	Group A	Group B	Total	Chi square value, df, p value
Male	11 (36.7%)	13 (43.3%)	24 (40%)	0.28, 1, 0.598
Female	19 (63.3%)	17 (56.7%)	36 (60%)	
Total	30 (100%)	30 (100%)	60 (100%)	

In the group A 19 (63.3%) participants were females and in the group B 17 (56.7%) participants were females.

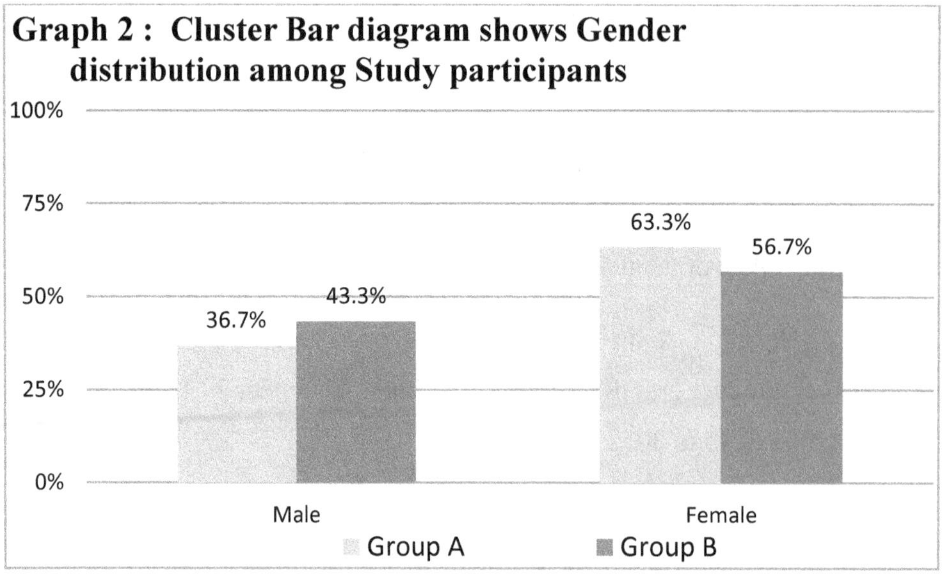

DIAGNOSIS PREDOMINANCE

Table 3: Diagnosis distribution among Study participants

Diagnosis	Group A	Group B	Total	Chi square value, df, p value
Acute Appendicitis	20 (66.7%)	23 (76.7%)	43 (71.7%)	0.74, 1, 0.390
Sub-Acute Appendicitis	10 (33.3%)	7 (23.3%)	17 (28.3%)	
Total	30 (100%)	30 (100%)	60 (100%)	

Graph 3: Cluster Bar diagram shows diagnosis distribution among Study participants

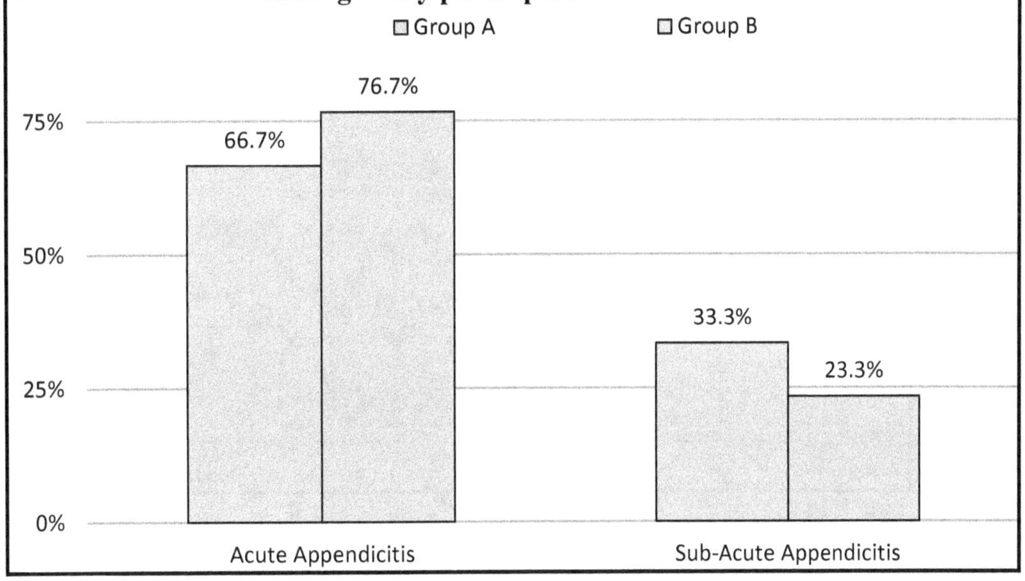

Out of 30 patients in group A, 20 (66.7%) patients had acute appendicitis and 10 (33.3%) patients had Sub-Acute Appendicitis. Out of 30 patients in group B, 23 (76.7%) patients had acute appendicitis and 7 (23.3%) patients had Sub-Acute Appendicitis. There is no significant difference seen between diagnosis distributions.

RELATIONSHIP BETWEEN AGE AND DIAGNOSIS

Table 4: Comparison between diagnosis distribution and age category

Diagnosis	Acute Appendicitis	Sub - Acute Appendicitis	Total	Chi square value, df, p value
15 to 30 Years	31 (72.1%)	9 (52.9%)	40 (66.7%)	2.01, 1, 0.156
> 31 to 60 years	12 (27.9%)	8 (47.1%)	20 (33.3%)	
Total	43 (100%)	17(100%)	60 (100%)	

Graph 4 : Cluster Bar diagram shows comparison between diagnosis distribution with age category

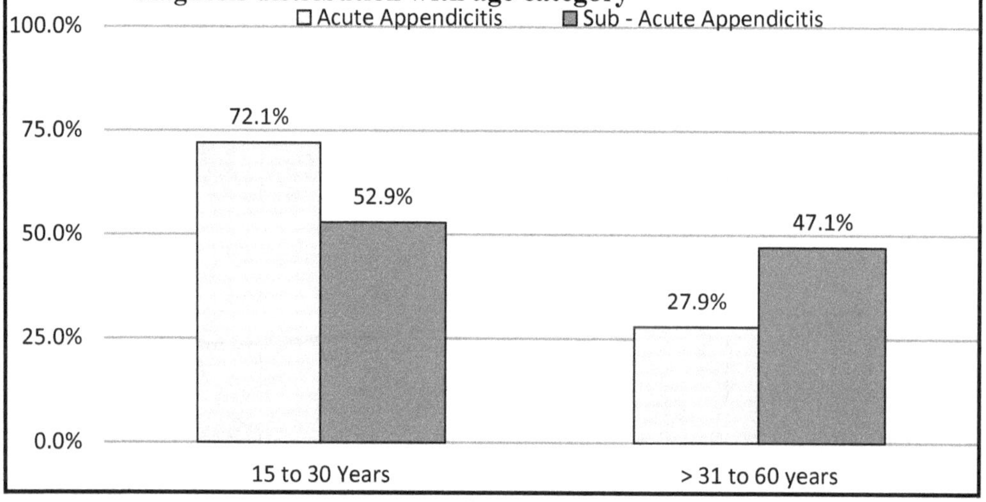

Among 43 patients in acute appendicitis, 31 (72.1%) patients were in the age category of 15 to 30 years and 12 (27.9%) patients were in age category of > 31 to 60 years. Out of 17 patients in Sub-acute appendicitis, 9 (52.9%) patients were in the age category of 15 to 30 years and 8 (47.1%) patients were in age category of > 31 to 60 years. Acute appendicitis was more common in the age category of 15 to 30 years when compared to >31 to 60 years of age and it is not significant.

SURGICAL PREDOMINANCE

Table 5: Surgery performed between Group A and Group B

Surgery performed	Group A	Group B	Total	Chi square value, df, p value
Laparoscopic appendectomy	9 (30.0%)	8 (26.7%)	17 (28.3%)	0.82, 1, 0.774
Open appendectomy	21 (70.0%)	22 (73.3%)	43 (71.7%)	
Total	30 (100%)	30 (100%)	60 (100%)	

Graph 5: Cluster Bar diagram shows diagnosis distribution among Study participants

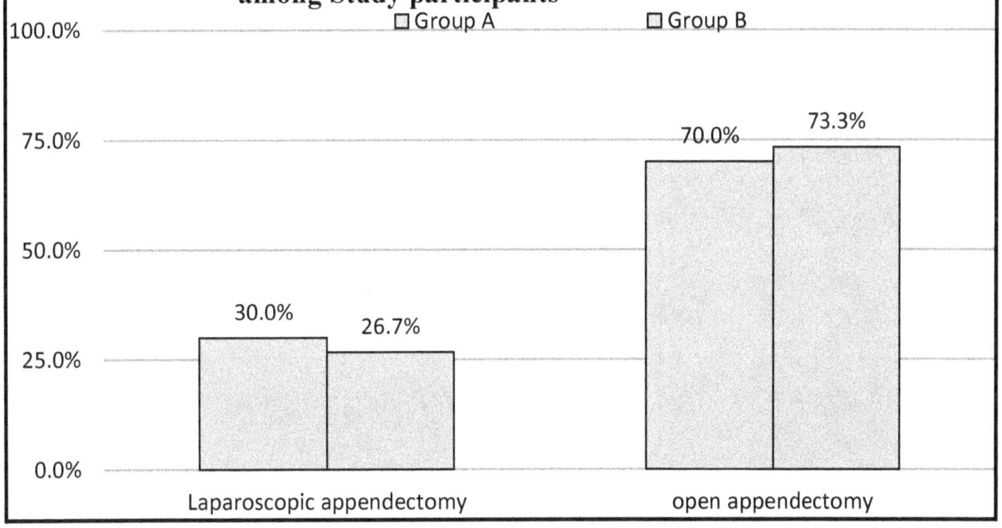

Out of 30 patients in group A, 9 (30.0%) patients underwent Laparoscopic appendectomy and 21 (70.0%) patients underwent Open appendectomy. Among 30 patients in group B, 8 (26.7%) patients underwent Laparoscopic appendectomy and 22 (73.3%) patients underwent Open appendectomy.

DURATION OF SURGERY

Table 6: Mean duration of surgery distribution between two groups

Group	N	Mean duration of surgery in hours	$\pm$ Std. Deviation	T value	P value
Group A	30	2.09	$\pm$ 0.65	0.20	0.524
Group B	30	1.99	$\pm$ 0.67		

Note: P value based on Independent sample t test

Graph 6 : Error bar diagram shows Duration of surgery distribution among study participants.

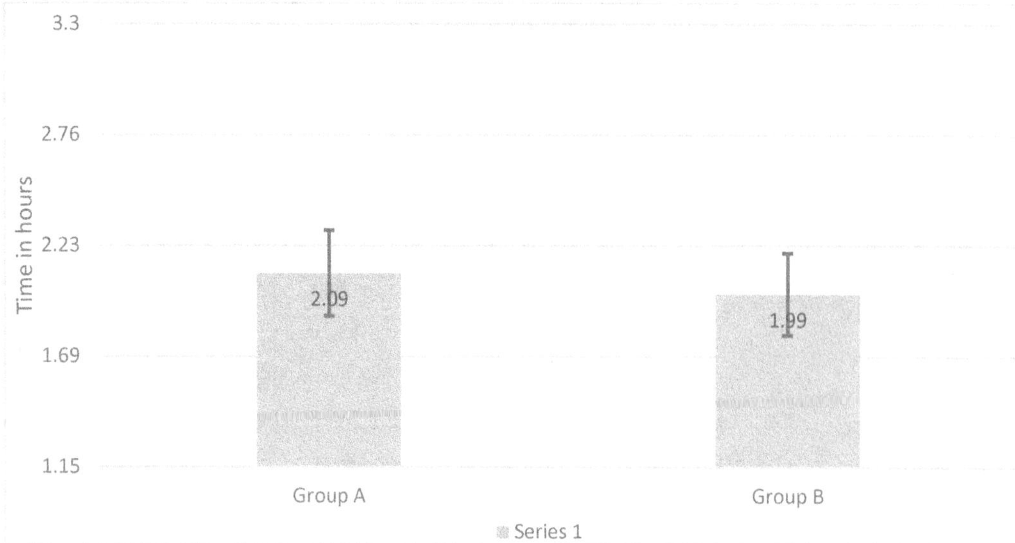

The mean duration of surgery among group A was 2.09 ± 0.65 hours and in group B was 1.99 ± 0.67 hours. The mean duration of surgery was equally distributed in two groups.

APPEARANCE OF FIRST BOWEL SOUND

Table 7: Mean time of first bowel sound post operatively between two groups

Group	N	Mean time of first bowel sound post operatively in hours	± Std. Deviation	T value	P value
Group A	30	6.57	± 1.89	0.35	0.001
Group B	30	9.67	± 2.21		

Note: P value based on Independent sample t test

Graph 7 : Error bar diagram shows Mean time of first bowel sound post operatively among study participants.

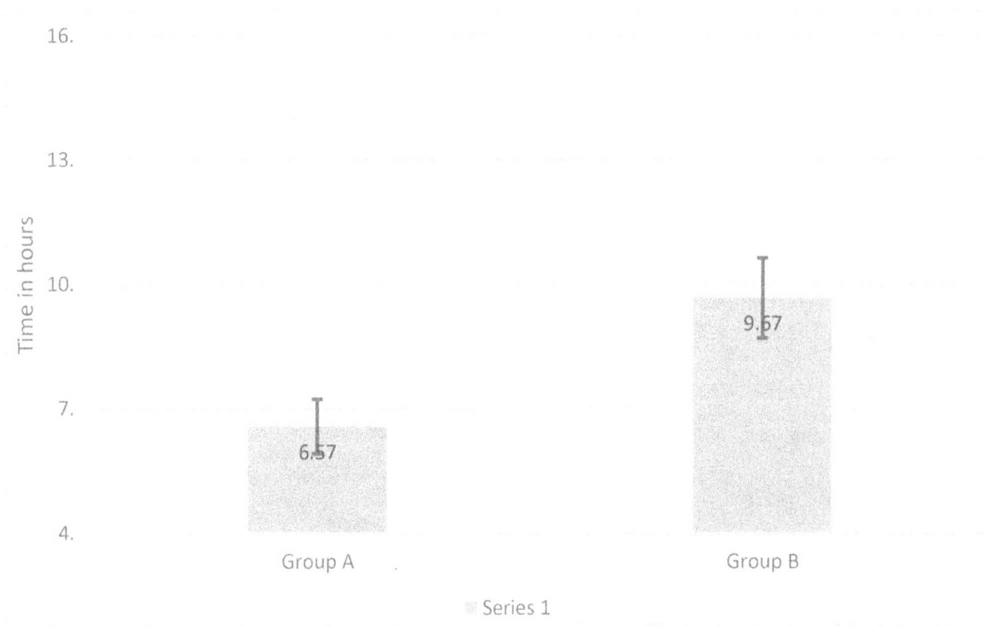

The mean time of first bowel sound felt in post-operative period among group A was 6.57 ± 1.89 hours and in group B was 9.67 ± 2.21 hours. The mean time of first bowel sound felt was significantly lower in group A when compared to group B with the p value of less than 0.05.

PASSAGE OF FIRST FLATUS

Table 8: Mean time of passing first flatus post operatively between two groups

Group	N	Mean time of first bowel sound post operatively in hours	± Std. Deviation	T value	P value
Group A	30	12.67	± 2.73	0.51	0.001
Group B	30	17.20	± 2.45		

Note: P value based on Independent sample t test

Graph 8 : Error bar diagram shows Mean time of passing first flatus post operatively among study participants.

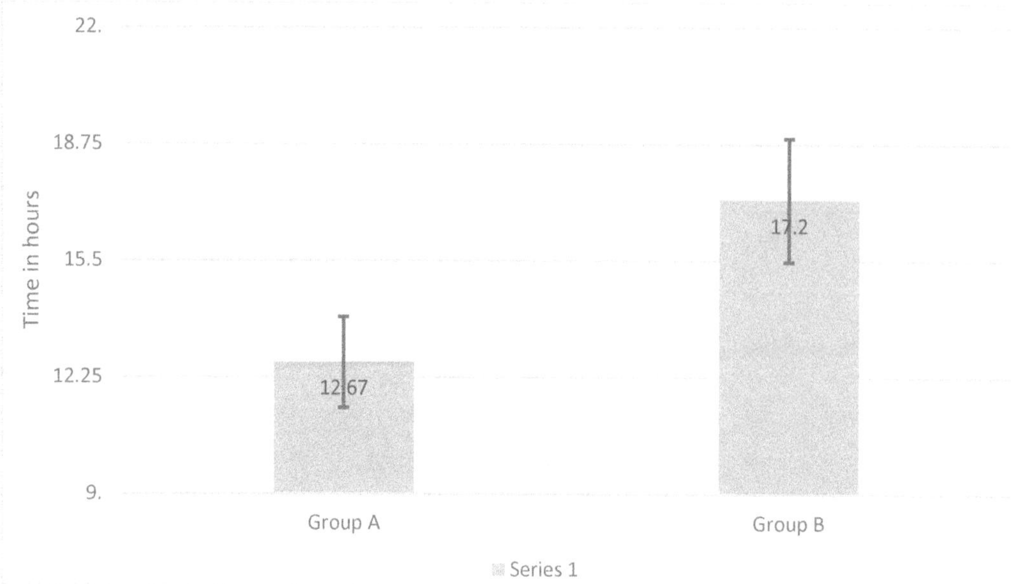

The mean time of passing first flatus in post-operative period among group A was 12.67 ± 2.73 hours and in group B was 17.20 ± 2.45 hours. The mean time of passing first flatus was significantly lower in group A when compared to group B with the p value of less than 0.05.

PASSAGE OF FIRST STOOLS

Table 9: Mean time of passing first stool post operatively between two groups

Group	N	Mean time of passing first stool post operatively in hours	± Std. Deviation	T value	P value
Group A	30	18.30	± 2.91	8.92	0.001
Group B	30	26.40	± 2.26		

Note: P value based on Independent sample t test

Graph 9 : Error bar diagram shows Mean time of passing first stool post operatively among study participants.

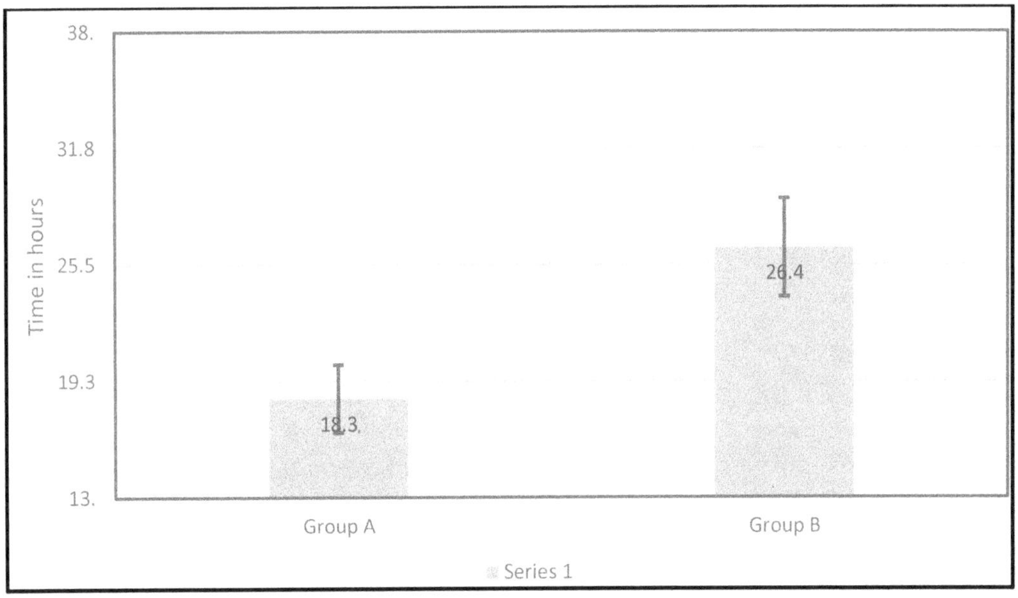

The mean time of passing first stool in post-operative period among group A was 18.30 ± 2.91 hours and in group B was 26.40 ± 2.26 hours. The mean time of passing first stool was significantly lower in group A when compared to group B with the p value of less than 0.05.

POSTOPERATIVE COMPLICATIONS

Table 10: Post-operative complaints among Study participants

Sex Distribution	Group A	Group B	Total	Chi square value, df, p value
No complaints	26 (86.7%)	27 (90.0%)	53 (88.3%)	0.16, 1, 0.688
Nausea / Vomiting	4 (13.3%)	3 (10.0%)	7 (11.7%)	
Total	30 (100%)	30 (100%)	60 (100%)	

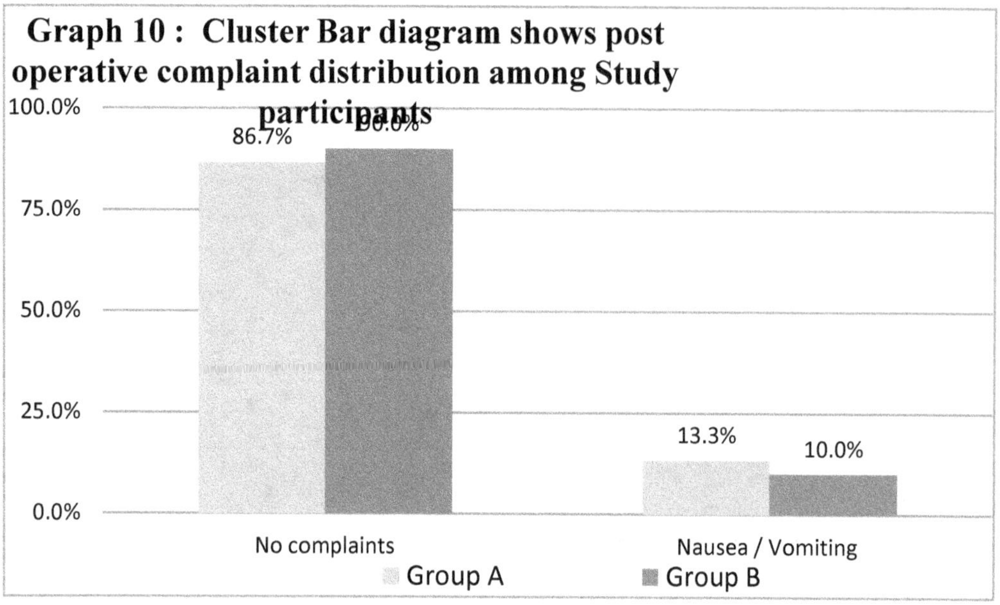

Graph 10 : Cluster Bar diagram shows post operative complaint distribution among Study participants

Among 30 patients in group A, 4 (13.3%) had post-operative complaints of Nausea/Vomiting and 26 (86.7%) patients don't have any complaints. Out of 30 patients in group B. 3 (10.0%) had post-operative complaints of Nausea/Vomiting and 27 (90.0%) patients don't have any post-operative complaints.

DISCUSSION

The current study set out to determine how chewing gum after a appendicectomy affected the recovery of intestinal function. The study's conclusions showed t hat both groups' intestinal parameters differed statistically significantly from one another.

According to the study's findings, neither group's age group, educational level, employment status, nor place of residence showed any statistically significant differences. Likewise, neither group's obstetrics variable showed any statistically significant differences, indicating that both groups were well matched.

The trial was completed by all 300 subjects. Both groups shared similar demographic characteristics, such as age, BMI, and parity.

The difference in the other post-operative statistics for the two groups, such as operating time and the number of blood transfusions, was not statistically significant.

Not procedures truly affect the gastro-intestinal tract, and depending on the surgery, recovering vary. Contrarily, colorectal surgery modifies and influences the digestive tract's integrity.

Gum chewing may have modest effects on lowering some difficulties, but the restoration. Chewing gum, however, clearly plays a part in boosting gastrointestinal motility in surgical procedures where the digestive tract is still in tact.

However, it is still unknown how chewing gum actually works. The theory is that chewing gum simulates the process of eating, which may greatly increase.

Gum chewing activates the cephalic - vagal axis, which in turn stimulates intestine myoelectric activity to counteract the activation of gastro-intestinal opioid receptors, according to the postulated physiological mechanism.

The potential contribution of of intestinal function has attracted a lot of attention recently. RCT data, however, do not yet support a conclusion.

The effectiveness of has been subject of published meta-analyses. Adults' postoperative ileus is affected by chewing gum, according. According to the study, chewing gum helps shorten the time it takes for the first bowel movement and flatus to pass.

Indicated was safe and did improve GI healing after colorectal surgery. This study found a relationship between chewing gum and a shorter contrary to our findings.

Sugarless gum was linked to better results after elective intestinal resection, according to a different that examined 9 relevant trials. However, there was no advantage o n the effectiveness of a lower rate of clinical problems or lower cost. This is consistent with our study's findings as well.

The findings of the current study, however, differ from those of a study conducted in Thailand by Jakkaew & Charoenkwan (2013) about the impact of chewing gum on the timing of the recovery of bowel function. Although it was noted that the group that chewed gum had a shorter median time before their first flatus, other intestinal measures did not show a statistically significant difference.

Additionally, a study reported on 67 CS patients who were divided into three groups and given general anesthesia that the chewing gum group had women who had earlier bowel sounds than the other groups.

Several RCTs examining the effectiveness with moderate to large sample sizes have recently been published. There have also been numerous publications. However, a number of these research' weaknesses need to be addressed.

The significant heterogeneity of the included studies limits the usefulness of these meta-analyses. Additionally, insufficient data was collected to do a subgroup analysis of the various interventions (such as the frequency and initiation of gum chewing).

There was no blinding done. Most studies did not use blinding, which had a substantial impact on could be challenging in this situation is a workable solution that might have been used to lessen detection bias. This might have improved the validity of the results.

Chewing gum usage decreases the risk of POI in appendicectomy patients, but it has no impact on how long they must stay in the hospital.

However, several studies came to a different conclusion. Researchers from the gum chewing was once again emphasized as a secure and straightforward preventive intervention to lower POI.

In an effort to explain the ambiguous findings from numerous research, we looked at potential contributing factors.

No trials were used in the investigations. Many of the measured differences could be accounted for by the bias caused by a lack of blinding. The groups weren't uniform.

The comparison of various surgical procedures, including laparoscopic and open surgery, was not warranted because it could lead to inter-group variability and muddle the data.

The different schedules for feeding could rationally change the outcomes. I n numerous t rials, patients were permitted to advance t o diet. Naturally, route feeding would b e m ore effective have an impact on the study's results.

Numbers and power may have contributed to the differences. These trials had a limited sample size, which could have resulted in a wide range of treatment effects owing to chance.

Following gastrointestinal surgery, the outcomes of randomized controlled trials were distinct from those of other abdominal surgeries. It was implied that this was because gastrointestinal tract involvement was not a factor in abdominal surgery, with the exception of gastrointestinal surgery.

Through a number of mechanisms, facilitates quicker restoration of GI function. The integrity of the gastrointestinal tract was changed by surgery. The integrity of the digestive system was crucial had only a little additive effect.

Improves the recovery of regular bowel motility. Chewing gum can help with this directly, or it can help indirectly by lowering postoperative pain and opiate consumption.

Chewing gum is a cheap and beneficial intervention. According to the postoperative patients we evaluated, it seems to be a reasonably safe drug. Early postoperatively, it was well tolerated, and the level of patient satisfaction was high. Chewing gum as an additional treatment in postoperative care strongly implies a positive outcome.

The majority of studies have confirmed that chewing gum helped patients recover their bowel function more quickly after abdominal surgery, with the exception of gastrointestinal surgery. And adding chewing gum to the increased post-surgery rehabilitation program could be a successful strategy. Chewing gum has become an unique and easy method for assisting the restoration of gastrointestinal function.

Gum chewing's impact during gastrointestinal surgery was debatable. The surgical management of the digestive tract, which significantly impacted the recovery of gastrointestinal function, may be the primary cause. Therefore, it becomes challenging in the recovery from bowel surgery has to be further supported by rigorous homogenous, prospective, and blinded randomized controlled studies.

Therefore, after abdominal procedures, especially caesarean sections, patients should be encouraged to chew gum postoperatively.

CONCLUSION

The conclusion of this study is that though we usually follow standard postoperative protocols and wait for the gastro-intestinal motility to be achieved spontaneously , SHAM feeding using Chewing gum merits routine use so that we can achieve early entric system activity and therefore decreasing the usual wait period of oral feeds. This in turn reduces the postoperative stay of the patient and enhances early rehabilitation post surgery.

LIMITATIONS

The study included only a small sample size a from a single hospital, and hence generalisation or results cannot be done, and it requires evidence from other large similar studies for generalisation of results using sham feeding through chewing gum as a standard postoperative protocol . The drawback would be the patient cooperation for chewing gum .

RECOMMENDATIONS

This study recommends further studies to assess the efficacy of Chewing gum SHAM feeding along with regular standard postoperative care so as to prevent post operative ileus and provide better post operative rehabilitation for patients who have underwent abdominal surgery .

BIBLIOGRAPHY

[1] Guo SF, Padmadas SS, Zhao FM, Brown JJ, Stones RW. Delivery settings and caesarean section rates in China. Bull World Health Organ 2007;85:755e62.

[2] Mattei P, Rombeau JL. Review of the pathophysiology and management of postoperative ileus. World J Surg 2006;30:1382e91.

[3] Iyer S, Saunders WB, Stemkowski S. Economic burden of postoperative ileus associated with colectomy in the United States. J Manage Care Pharm 2009;15:485e94.

[4] Abd-El-Maeboud K, Ibrahim M, Shalaby D, Fikry M. Gum chewing stimulates early return of bowel motility after caesarean section. BJOG 2009;116: 1334e9.

[5] Kafali H, Duvan CI, G€ozdemir E, Simavli S, Onaran Y, Keskin E. Influence of gum chewing on postoperative bowel activity after cesarean section. Gynecol Obstet Invest 2010;69:84e7.

[6] Shang HK, Yang Y, Tong XW, Zhang LJ, Fang AM, Hong L. Gum chewing slightly enhances early recovery from postoperative ileus after cesarean section: results of a prospective, randomized, controlled Trial. Am J Perinatol 2010;2: 387e91.

[7] Ledari FM, Barat S, Delavar MA. Chewing gums has stimulatory effects on bowel function in patients undergoing cesarean section: A randomized controlled trial. Bosn J Basic Med Sci 2012;12:265e8.

[8] Jakkaew B, Charoenkwan K. Effects of gum chewing on recovery of bowel function following cesarean section: a randomized controlled trial. Arch Gynecol Obstet 2013;288:255e60.

[9] Purkayastha S, Tilney HS, Darzi AW, Tekkis PP. Meta-analysis of randomized studies evaluating chewing gum to enhance postoperative recovery following colectomy. Arch Surg 2008;143:788e93.

[10] Fitzgerald JE, Ahmed I. Systematic review and meta-analysis of chewing-gum therapy in the reduction of postoperative paralytic ileus following gastrointestinal surgery. World J Surg 2009;33:2557e66.

[11] Yin Z, Sun J, Liu T, Zhu Y, Peng S, Wang J. Gum chewing: another simple potential method for more rapid improvement of postoperative gastrointestinal function. Digestion 2013;87:67e74.

[12] Li S, Liu Y, Peng Q, Xie L, Wang J, Qin X. Chewing gum reduces postoperative ileus following abdominal surgery: a meta-analysis of 17 randomized controlled trials. J Gastroenterol Hepatol 2013;28:1122e32.

[13] Moher D, Liberati A, Tetzlaff J, Altman DG. PRISMA Group. Preferred reporting items for systematic reviews and meta-analyses: the PRISMA statement. Ann Intern Med 2009;151:264e9.

[14] Higgins JP, Green S. Cochrane Handbook for Systematic Reviews of Interventions version 5.1.0 (Updated on 2011). Available at: http://handbook. cochrane.org/. [accessed 25.05.14].

[15] Higgins JP, Altman DG, Gøtzsche PC, Jüni P, Moher D, Oxman AD, et al. The Cochrane Collaboration's tool for assessing risk of bias in randomized trials. BMJ 2010;343:d5928.

[16] Higgins JP, Thompson SG, Deeks JJ, Altman DG. Measuring inconsistency in meta- analyses. BMJ 2003;327:557e60.

[17] Hozo SP, Djulbegovic B, Hozo I. Estimating the mean and variance from the median, range, and the size of a sample. BMC Med Res Methodol 2005;5:13.

[18] Ledari FM, Barat S, Delavar MA, Banihosini SZ, Khafri S. Chewing sugar-free gum reduces ileus after cesarean section in nulliparous women: a randomized clinical trial. Iran Red Crescent Med J 2013;15:330e4.

[19] Byrne H. Gum chewing stimulates early return of bowel motility after caesarean section. BJOG 2010;117:117.

[20] Keenahan M. Does gum chewing prevent postoperative paralytic ileus? Nursing 2014;44:1e2.

[21] Abd Rabbo S. Early oral hydration: a novel regimen for management after elective cesarean section. J Obstet Gynaecol (Tokyo 1995) 1995;21:563e7.

[22] Traut U, Brügger L, Kunz R, Pauli-Magnus C, Haug K, Bucher HC, et al. Systemic prokinetic pharmacologic treatment for postoperative adynamic ileus following abdominal surgery in adults. Cochrane Database of Systematic Reviews 2008;23:CD004930.

[23] Charoenkwan K, Palapinyo C. Early solid food after cesarean section and postoperative ileus. Int J Gynaecol Obstet 2005;90:144e5.

[24] Johnson MD, Walsh RM. Current therapies to shorten postoperative ileus. Cleve Clin J Med 2009;76:641e8.

[25] Fannin J Valea FA. (2011). Perioperative bowel management for gynecologic surgery. A Obstet Gynecol; 205: 309-314.

[26] Ho Y. M., Smith S. R., Pockney P., Lim P., and Attia J. (2014). "A meta-analysis on the effect of sham feeding following colectomy: should gum chewing be included in enhanced recovery after surgery protocols?" Diseases of the Colon and Rectum, vol. 57, no. 1, pp. 115-26.

[27] Keller D., Stein SL. (2013). Facilitating return of bowel function after colorectal surgery: alvimopan and gum chewing. Clin Colon Rectal Surg. Sep; 26 (3): 186-90.

[28] Brooke B., Goodney P., & Kraiss L. (2016). Readmission Destination and Risk of Mortality After Major Surgery: An Observational Cohort Study. Journal of Vascular Surgery, 63 (4), 1126.

[29] Kaplan J. R., Lee Z., Eun D. D., & Reese A. C. (2016). Complications of Minimally Invasive Surgery and Their Management. Current Urology Reports, 17(6).

[30] Tan W. J., Chew M. H., Dharmawan A. R. (2016). "Critical appraisal of laparoscopic vs open rectal cancer surgery," World Journal of Gastrointestinal Surgery, vol. 8, no. 6, pp. 452-60.

[31] Barletta JF, Senagore AJ. (2014). Reducing the burden of postoperative ileus: evaluating and implementing an evidence-based strategy. World J Surg; 38: 1966-77.

www.ingramcontent.com/pod-product-compliance
Lightning Source LLC
LaVergne TN
LVHW080527210926
848481LV00022B/3640